THE PANAMA CANAL

Pictorial View of the
World's Greatest Engineering Feat

Linking the Atlantic and Pacific Oceans

With a Brief History and Description of the
Gigantic Undertaking

BY
THOMAS H. RUSSELL, A. M., LL. D.
Member National Geographic Society

THE HAMMING PUBLISHING CO.
CHICAGO, ILLINOIS

THE PANAMA CANAL

THE Panama Canal will soon be opened for navigation, and the passage of the first vessel from ocean to ocean through its hospitable locks and placid waters will signalize the successful completion of the greatest engineering feat in the world's history. Naturally enough, it will afford an occasion for pardonable pride and patriotic enthusiasm to every citizen of the United States, whose government has carried through this gigantic undertaking in spite of almost insuperable difficulties.

While the official date of opening the Canal has been set for January 1, 1915, it is the intention to allow vessels to utilize the new waterway just as soon as practicable. Present indications seem to bear out the opinion expressed by Colonel George W. Goethals, U. S. A., chairman and chief engineer of the Isthmian Canal Commission, that this can be accomplished during the latter half of 1913. Shipping interests all over the world will be advised as soon as the Commission feels assured that vessels can be passed without unnecessary delay.

The Panama Canal is indeed a tremendous work, dwarfing all other undertakings of human skill and labor. It links the oceans that have been hitherto communicable one with the other only at the expense of weeks of time and a mint of money. Time is the essence of all business movements nowadays—now more than ever before—and the time that will be saved daily in transit between the great trading centers by the opening of the Canal is of almost inestimable value.

In a recent message to Congress the President of the United States said that the first passage of ships through the Canal would mark an important era, not in the history of this country alone, but in that of the civilized world; and it will undoubtedly make a tremendous difference in the attitude of the nations that are commonly regarded as constituting the backbone of civilization.

When it is possible, for instance, to send from New York to San Francisco a great battleship or a fleet for the protection of our Western coast, in a small fraction of the time it used to take, that alone constitutes a marvelous advance in our methods and means of self-defense. And that is the kind of advance in martial facility that appeals to other nations and influences their attitude,—nations that may possibly contemplate the angles of attack upon the United States,—not now, let us hope and believe, but at some time in the future. Statesmen of the elder variety always figure ahead, and the defensive, non-aggressive nation must always prepare its bulwarks of defense, just as the offensive nation prepares its artillery.

The United States, seeking trouble with none, but offering a means of facile interoceanic transit to all, asks simply that it may enjoy the fruits of its own enterprise—and no self-respecting nation would ask less. Suppose the conditions were changed. Suppose some European nation, for instance, had built the great passageway between two oceans. Does anyone imagine for a single instant that that nation would surrender one jot or tittle of its right to control the Canal? Suppose France had succeeded in putting through a world's highway for commerce from the Atlantic to the Pacific, across Central American territory. Does anyone suppose for a moment that France, or Germany, or any other nation that had thus opened up a waterway of universal interest would not have retained the utmost possible benefit from its operation?

And why should the United States surrender one bit of the advantage it has gained by taking over from another power the privilege—for it is an international privilege—of building this great and wonderful link between the mighty oceans of the western world?

Soon to be opened, as the result of American enterprise and ingenuity, after failure had marked the attempts of the greatest European canal builders,—the Panama Canal will stand as an everlasting monument to American genius, the genius of construction that leads the world today; that erects gigantic structures in all the countries that have faith and capital to invest in modern methods; the constructive genius that knows no difficulties, that could reconstruct the Pyramids or the hanging gardens of Babylon, and that would make a night's work of building a second Sphinx, if the utility of the object could be demonstrated.

It is this constructive genius that is immortalized in the Panama Canal. America can build anything the world wants, and has the men and the money to do it—and the Old World probably needed the lesson. It needed to be taught that there are here in the United States the engineers to plan, the builders and mechanics to construct, and the ability and energy to carry out great enterprises that the Old World dreams about decade after decade and knows not how to realize.

Four hundred years had passed since first the idea of a passage from the Atlantic to the Pacific across the Isthmus that separates the north and south American continents first entered the head of civilized man. There were four centuries of talk about it—and little done until a practical, level-headed American occupying the Presidential chair of the United States saw not only the opportunity, but the feasibility of a definite plan—and was possessed of sufficient American initiative to go ahead with the work, even though it involved the recognition of a young Republic before any of the older sister nations had had their say.

So the United States stepped in, made the Republic of Panama possible, and secured the Canal Zone in which we proceeded to build the big waterway that will make our possessions more and more valuable as the years elapse.

Now the time approaches when we are to see the solid success of our endeavors; when ships shall pass from ocean to ocean under the American flag, and all the world shall pay us tribute for our daring and our skill. We have succeeded where others have failed; but in our triumph there is no note of undue exultation, no idea of boasting of our accomplishment;—simply a perfectly proper expression of pride in our achievement, a pride we have a right to feel as a nation of doers as well as thinkers; and we extend with the expression of our delight in achievement a cordial invitation to the nations of the world to come and enjoy with us the fruits of our high endeavor.

In building the Panama Canal we know that we have conferred a favor on humanity. We have helped to annihilate distance. We have aided in bringing the nations closer together. We have made trade and commerce easier and more profitable for all. We have wrought not for ourselves alone, but for civilization—and we present the result in an accomplished fact where others theorized and fell down.

We dedicate our work for the use and benefit of our own and succeeding generations. The first ship that passes through the Panama Canal, even though it be an American battleship, will be freighted with a message of peace and good will to all the earth—and they who cannot read that message aright deserve none of the benefits that will flow from the opening of the waterway.

In this book there is presented a pictorial review of the great Canal in the process of building. Completed, the Canal will be a picture. In the making it was not always pretty and the workers toiled and dug in the sweat of their brow. It is believed that none can view these pictures without being impressed with the magnitude of the work, with its difficulties and its dangers, and with the marvelous results of the engineering and administrative skill displayed by the men of the United States Army who finally, after many had tried and been found wanting, carried this mammoth undertaking to a successful conclusion.

HISTORY OF THE CANAL.

The history of the Isthmian Canal is a remarkable record of persistent human endeavor, covering four centuries of time, marked by many failures, and now at last about to be crowned with success.

The project has long been recognized as "an indispensable factor in the future of the American continent." Spain, England, Portugal, and France have all embarked upon the work, either directly or by giving aid and encouragement to their representatives, and failed. The time for success had not yet arrived, for even if the funds with which to prosecute the work had been unlimited, the difficulties were then too great for engineering and medical science to solve.

It was President Grant who first advanced the policy of "an American canal under American control." President Roosevelt, voicing the sentiments

Photo by Underwood & Underwood, N. Y.
Col. George W. Goethals, U. S. A.
Chief Engineer and Chairman Isthmian Canal Commission

of the entire American people, lent the aid of the United States in undertaking the work, and President Taft enthusiastically advanced the project, which will be completed, in all human probability, under the administration of President Wilson.

The idea of a canal across the Isthmus of Panama was born in the days of Balboa, who crossed the isthmus in 1513, but the project met with opposition in Spain under King Philip II., and was laid on the shelf for two centuries or more. In 1814 it was revived, but by that time Spain had lost her Central and South American colonies, and ceased to be a factor in canal affairs.

England investigated Isthmian conditions with reference to a canal early in the nineteenth century, through Lord Nelson and Baron von Humboldt, but nothing practical resulted from their reports. It is interesting to note that Goethe about this time prophesied an Isthmian canal under American control.

In 1835 the United States first became interested in the project through a resolution introduced in the Senate by Henry Clay, but the panic of 1837 effectually estopped action contemplated at that time.

In 1838 a concession was granted to a French company for the construction of highways, railroads or a canal across the Isthmus, but the concession lapsed for lack of capital.

The year 1855 saw the opening of the Panama Railroad, constructed by Americans across the Isthmus. The concession of the company gave it control of the Panama route for a canal, but with railroad communication firmly established and financially successful, the idea of canal construction was relegated to the background.

Meanwhile other canal routes were exploited by a small army of promoters. Altogether nineteen different routes have been suggested and received more or less attention. Of these, the Tehuantepec, Nicaragua, Panama, and Darien projects are the most important, and Nicaragua has been Panama's principal rival in the last thirty years.

In 1869 the United States again took up the canal question, and President Grant appointed an interoceanic canal commission. But nothing definite was done and France stepped into the arena in 1876, and remained in control of operations for twenty-eight years, until 1904, when the French retired, defeated, in favor of the United States. This included the period of effort of the great French engineer, Ferdinand de Lesseps, builder of the Suez Canal. His unsuccessful aim was to build a sea-level canal across the Isthmus.

Progress having ceased at Panama under the second French Canal Company, on March 3, 1899, the Congress of the United States passed an act authorizing the President to make full and complete investigations of the Isthmus of Panama with a view to the construction of a canal to connect the Atlantic and Pacific oceans.

This marks the opening of the last chapter in the construction of the Panama Canal, the end of which is now, by the early completion of the canal, in sight. The commission appointed in accordance with the above act was called upon to investigate particularly the Nicaragua and the Panama routes and to report which was the more practicable and feasible, and the cost. In November, 1901, it reported in favor of the Nicaragua route. The price fixed by the Panama Canal Company was $109,000,000. By subsequent negotiations the French company was induced to reduce its price to $40,000,000, and the commission in January, 1902, submitted a supplemental report in favor of the Panama route.

Satisfactory arrangements were completed for the purchase of the French company's rights, etc., for $40,000,000, and negotiations with the Republic of Colombia were carried on to secure other necessary rights and privileges not held by the French company. After a long delay, a satisfactory treaty was formulated, which was rejected by Colombia in 1903.

The province of Panama, an integral part of Colombia, thereupon seceded and organized an independent republic, with an area of about 31,000 square miles and a population which at present is stated to be 419,000. This resulted in the negotiation of a satisfactory treaty with the new Republic of Panama, including the payment, under certain terms, of $10,000,000 by the United States to the Republic of Panama and an annual payment of $250,000 beginning nine years after the signing of the treaty. Under this treaty the United States guaranteed the independence of the Republic of Panama and secured absolute control over what is now called the Canal Zone, a strip of land about 10 miles in width, with the canal through the center, and 45 miles in length from sea to sea, with an area of about 448 square miles. The United States also has jurisdiction over the adjacent water for three miles from shore.

The act of Congress of 1902 placed entire jurisdiction in regard to the construction of the canal in the hands of the President of the United States, the particular functions in regard thereto being exercised by a commission of seven members. For convenience in administration the canal operations were placed under the Secretary of War.

The formal transfer of the property of the French Canal Company to the United States took place on May 4, 1904, and the first two and one-half years thereafter, or until January, 1907, were devoted largely to the work of preparation.

Meanwhile the lock type of canal had been decided upon, and on June 29, 1906, its construction was authorized by Congress, and promptly entered upon by the Isthmian Canal Commission, which, with Col. George W. Goethals, U. S. A., as chairman and chief engineer, will soon see the full fruition of its splendid patriotic endeavors.　　T. H. R.

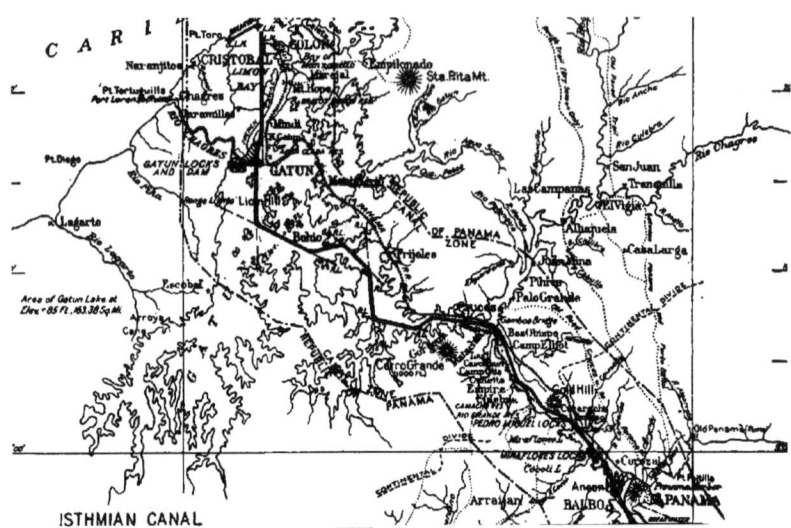

Map Showing Isthmus with Completed Canal.

PLAN OF THE CANAL

The entire length of the Panama Canal from deep water in the Atlantic to deep water in the Pacific is about 50 miles. Its length from shore-line to shore-line is about 40 miles. In passing through it from the Atlantic to the Pacific, a vessel will enter the approach channel in Limon Bay, which will have a bottom width of 500 feet and extend to Gatun, a distance of about seven miles. At Gatun, it will enter a series of three locks in flight and be lifted 85 feet to the level of Gatun Lake. It may steam at full speed through this lake, in a channel varying from 1,000 to 500 feet in width, for a distance of about 24 miles, to Bas Obispo, where it will enter the Culebra Cut. It will pass through the Cut, a distance of about nine miles, in a channel with a bottom width of 300 feet, to Pedro Miguel. There it will enter a lock and be lowered 30⅓ feet to a small lake, at an elevation of 54⅔ feet above sea level, and will pass through this for about 1½ miles to Miraflores. There it will enter two locks in series and be lowered to sea level, passing out into the Pacific through a channel about 8½ miles in length, with a bottom width of 500 feet. The depth of the approach channel on the Atlantic side, where the maximum tidal oscillation is 2½ feet, will be 41 feet at mean tide, and on the Pacific side, where the maximum oscillation is 21 feet, the depth will be 45 feet at mean tide.

Throughout the first 16 miles from Gatun, the width of the Lake channel will be 1,000 feet; then for 4 miles it will be 800 feet, and for 4 miles more, to the northern entrance of Culebra Cut at Bas Obispo, it will be 500 feet. The depth will vary from 85 to 45 feet. The water level in the Cut will be that of the Lake, the depth 45 feet, and the bottom width of the channel 300 feet.

Three hundred feet is the minimum bottom width of the Canal. This width begins about half a mile above Pedro Miguel locks and extends about 8 miles through Culebra Cut, with the exception that at all angles the channel is widened sufficiently to allow a thousand-foot vessel to make the turn. The Cut has eight angles, or about one to every mile. The 300-foot widths are only on tangents between the turning basins at the angles. The smallest of these angles is 7° 36′, and the largest 30°.

GATUN DAM

The Gatun Dam, which will form Gatun Lake by impounding the waters of the Chagres and its tributaries, will be nearly 1½ miles long, measured on its crest, nearly ½ mile wide at its base, about 400 feet wide at the water surface, about 100 feet wide at the top, and its crest, as planned, will be at an elevation of 115 feet above mean sea level, or 30 feet above the normal level of the Lake. Of the total length of the Dam only 500 feet, or one-fifteenth, will be exposed to the maximum water head of 85 feet. The interior of the Dam is being formed of a natural mixture of sand and clay, dredged by hydraulic process from pits above and below the Dam, and placed between two large masses of rock and miscellaneous material obtained from steam shovel excavation at various points along the Canal. The top and upstream slope will be thoroughly riprapped. The entire Dam will contain about 21,000,000 cubic yards of material.

The Spillway is a concrete lined opening, 1,200 feet long and 300 feet wide, cut through a hill of rock nearly in the center of the Dam, the bottom of the opening being 10 feet above sea level. It will contain about 225,000 cubic yards of concrete. During the construction of the Dam, all the water discharged from the Chagres and its tributaries will flow through this opening. When construction has advanced sufficiently to permit the Lake to be formed, the Spillway will be closed with a concrete dam, fitted with gates and machinery for regulating the water level of the Lake.

WATER SUPPLY OF GATUN LAKE

Gatun Lake will impound the waters of a basin comprising 1,320 square miles. When the surface of the water is at 85 feet above sea level, the Lake will have an area of about 164 square miles, and will contain about 206 billion cubic feet of water. During eight or nine months of the year, the lake will be kept constantly full by the prevailing rains, and consequently a surplus will need to be stored for only three or four months of the dry season.

The smallest run-off of water in the basin, during the past 21 years, as measured at Gatun, was about 146 billion cubic feet. In 1910 the run-off was 360 billion cubic feet, or a sufficient quantity to fill the lake one and a half times. The water surface of the Lake will be maintained during the rainy season at 87 feet above sea level, making the minimum channel depth in the Canal 47 feet. As navigation can be carried on with about 41 feet of water, there will be stored for dry season surplus over five feet of water. Making due allowance for evaporation, seepage, leakage at the gates, and power consumption, this would be ample for 41 passages daily through the locks, using them at full length, or about 58 lockages a day when partial length is used, as would be usually the case, and when cross filling from one lock to the other through the central wall is employed. This would be a larger number of lockages than would be possible in a single day. The average number of lockages through the Sault Ste. Marie Canal on the American side was 37 per day in the season of navigation of 1909, which was about eight months long. The average number of ships passed was about 1½ per lockage. The freight carried was more than 30,000,000 tons. The Suez Canal passed about 12 vessels per day, with a total tonnage for the year of 15,500,000.

DAMS ON PACIFIC SIDE

The water level of Gatun Lake, extending through the Culebra Cut, will be maintained at the south end by an earth dam connecting the locks at Pedro Miguel with the high ground to the westward, about 1,400 feet long, with its crest at an elevation of 105 feet above mean tide. A concrete core wall, containing about 700 cubic yards, will connect the locks with the hills to the eastward; this core wall will rest directly on the rock surface and is designed to prevent percolation through the earth, the surface of which is above the Lake level.

A small lake between the locks at Pedro Miguel and Miraflores will be formed by dams connecting the walls of Miraflores locks with the high ground on either side. The dam to the westward will be of earth, about 2,700 feet long, having its crest about 15 feet above the water in Miraflores Lake.

The east dam will be of concrete, containing about 75,000 cubic yards; will be about 500 feet long, and will form a spillway for Miraflores Lake, with crest gates similar to those at the Spillway of the Gatun Dam.

THE LOCKS

There will be six double locks in the Canal; three pairs in flight at Gatun, with a combined lift of 85 feet; one pair at Pedro Miguel, with a lift of 30½ feet, and two pairs at Miraflores, with a combined lift of 54⅔ feet at mean tide. The usable dimensions of all are the same—a length of 1,000 feet, and width of 110 feet. Each lock will be a chamber, with walls and floor of concrete, and mitering gates at each end.

The side walls will be 45 to 50 feet wide at the surface of the floor; will be perpendicular on the face, and will narrow from a point 24⅓ feet above the floor until they are 8 feet wide at the top. The middle wall will be 60 feet wide, approximately 81 feet high, and each face will be vertical. At a point 42⅓ feet above the surface of the floor, and 15 feet above the top of the middle culvert, this wall will divide into two parts, leaving a space down the center much like the letter "U," which will be 19 feet wide at the bottom and 44 feet wide at the top. In this center space will be a tunnel divided into three stories, or galleries. The lowest gallery will be for drainage; the middle, for the wires that will carry the electric current to operate the gate and valve machinery installed in the center wall, and the upper will be a passageway for the operators.

The lock gates will be steel structures 7 feet thick, 65 feet long, and from 47 to 82 feet high. They will weigh from 300 to 600 tons each. Ninety-two leaves will be required for the entire Canal, the total weighing 57,000 tons. Intermediate gates will be used in the locks, in order to save water and time, if desired, in locking small vessels through, the gates being so placed as to divide the locks into chambers 600 and 400 feet long, respectively. Ninety-five per cent of the vessels navigating the high seas are less than 600 feet long. In the construction of the locks, it is estimated that there will be used approximately 4,200,000 cubic yards of concrete, requiring about the same number of barrels of cement.

Electricity will be used to tow all vessels into and through the locks, and to operate all gates and valves, power being generated by water turbines from the head created by Gatun Lake. Vessels will not be permitted to enter or pass through the locks under their own power, but will be towed through by electric locomotives running on cog-rails laid on the tops of the lock walls. There will be two towing tracks for each flight of locks, one on the side and one on the middle wall. On each side wall there will be one return track and on the middle wall a third common to both of the twin locks. All tracks will run continuously the entire length of the respective flights and will extend some distance on the guide approach walls at each end. The number of locomotives used will vary with the size of the vessel. The usual number required will be four; two ahead, one on each wall, imparting motion to the vessel, and two astern, one on each wall, to aid in keeping the vessel in a central position and to bring it to rest when entirely within the lock chamber. They will be equipped with a slip drum, towing windlass and hawser which will permit the towing line to be taken in or paid out without actual motion of the locomotive on the track.

The locks will be filled and emptied through a system of culverts. One culvert 254 sq. ft. in area of cross section, about the area of the Hudson River tunnels of the Pennsylvania Railroad, extends the entire length of each of the middle and side walls and from each of these large culverts there are several smaller culverts, 33 to 44 sq. ft. in area, which extend under the floor of the lock and communicate with the lock chamber through holes in the floor. The large culverts are controlled at points near the miter gates by large valves and each of the small culverts extending from the middle wall culvert into the twin chambers is controlled by a cylindrical valve. The large culvert in the middle wall feeds in both directions through laterals, thus permitting the passage of water from one twin lock to another, effecting a saving of water. (*See cuts.*)

To fill a lock the valves at the upper end are opened and the lower valves closed. The water

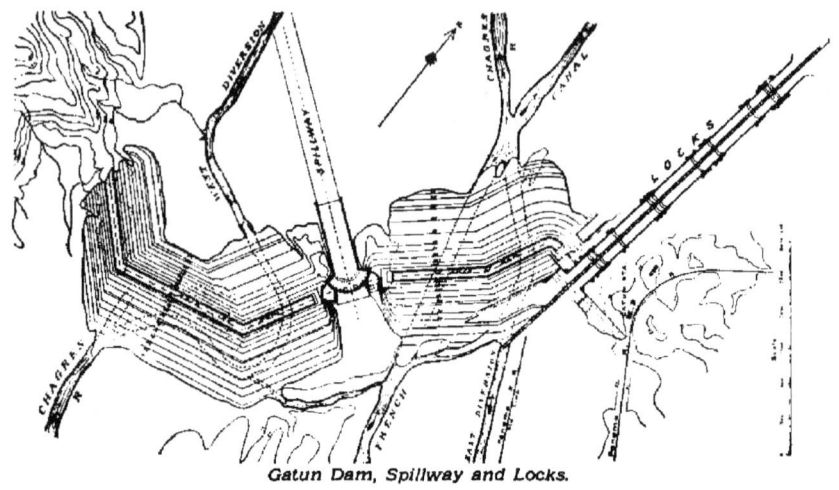

Gatun Dam, Spillway and Locks.

flows from the upper pool through the large culverts into the small lateral culverts and thence through the holes in the floor into the lock chamber. To empty a lock the valves at the upper end are closed and those at the lower end are opened and the water flows into the lower lock or pool in a similar manner. This system distributes the water as evenly as possible over the entire horizontal area of the lock and reduces the disturbance in the chamber when it is being filled or emptied.

The depth of water over the miter sills of the locks will be 40 feet in salt water and $41\frac{1}{3}$ feet in fresh water.

. The average time of filling and emptying a lock will be about fifteen minutes, without opening the valves so suddenly as to create disturbing currents in the locks or approaches. The time required to pass a vessel through all the locks is estimated at 3 hours; one hour and a half in the three locks at Gatun, and about the same time in the three locks on the Pacific side. The time of passage of a vessel through the entire Canal is estimated as ranging from 10 to 12 hours, according to the size of the ship, and the rate of speed at which it can travel

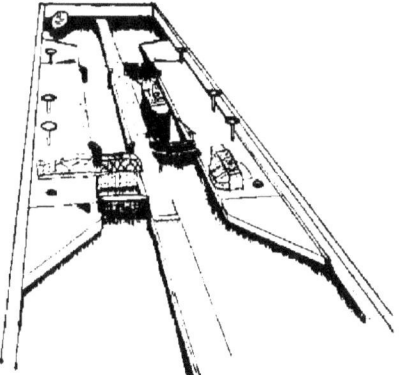

MODEL OF PEDRO MIGUEL LOCKS

The lock on the right is nearly filled for an upward lockage. Four electric locomotives are shown securely holding a 10,000-ton ship, and ready to tow it out of the lock, so soon as the upper gates are opened. In the foreground is shown a protective chain; at the entrance to the lock on the left is shown a caisson in position and acting as a barrier between the high level above and the low level below the lock.

On the right is shown an emergency dam in its normal position when not in use and on the left the other dam is shown swung in position across the lock with the wicket girder down in readiness to support the wickets or gates which complete the barrier.

SLIDES

There are in all twenty-one slides along the Culebra Cut. Twelve cover areas varying from one to forty-seven acres, and nine cover areas of less than one acre each, making in all a total of one hundred and forty-nine acres. The largest is the Cucaracha slide, on the east side of the Canal, which covers an area of forty-seven acres, and which has broken back 1,820 feet from the center line of the Canal. This slide, according to French records, started as early as 1884, and has given the Americans considerable trouble since they began work. Over two million cubic yards have been removed by the Americans, and the slide is still active. The next largest slide is a combination of two slides on the west side of the Cut at Culebra, just north of Contractor's Hill, covering about twenty-eight acres. Over two million cubic yards have been removed from this slide, and it is estimated that one million cubic yards are still in motion. On the east side of the Cut, north of Gold Hill, is another large slide covering an area of about seventeen acres which has broken back 1,200 feet from the center line of the Canal. Over 416,000 cubic yards have been taken out of this slide and about three-quarters of a million more are still in motion. The total distance across the Cut at this point from back to back of slides is 1,950 feet. In all, over nine million cubic yards have been taken out since July, 1905, because of slides, and over three million cubic yards are still in motion.

Side Wall of Locks Compared with Six-story Building.

CAPACITY OF STEAM SHOVELS AND DIRT TRAINS

There are several classes of steam shovels engaged in excavating work, equipped with dippers ranging in capacity from 1¾ cubic yards to 5 cubic yards, and a trenching shovel, which has a dipper with a capacity of ¾ of a cubic yard.

Each cubic yard, place measurement, of average rock weighs about 3,900 pounds; of earth, about 3,000 pounds; of "the run of the cut," about 3,600 pounds, and is said to represent about a two-horse cart load. Consequently, a five-cubic yard dipper, when full, carries 8.7 tons of rock, 6.7 tons of earth, and 8.03 tons of "the run of the cut."

Three classes of cars are used in hauling spoil— flat cars with one high side, which are unloaded by plows operated by a cable upon a winding drum, and two kinds of dump cars, one large and one small. The capacity of the flat cars is 19 cubic yards; that of the large dump cars, 17 cubic yards, and that of the small dump cars, 10 cubic yards. The flat car train is ordinarily composed of 20 cars in hauling from the cut at Pedro Miguel, and of

21 cars in hauling from the cut at Matachin. The large dump train is composed of 27 cars, and the small dump train of 35 cars.

The average load of a train of flat cars, in hauling the mixed material known as "the run of the cut," is 610.7 tons (based on a 20-car train) ; of a train of large dump cars, 737.68 tons, and of a train of small dumps, 562.5 tons.

The average time consumed in unloading a train of flat cars is from 7 to 15 minutes; in unloading a train of large dump cars, 15 to 40 minutes, and in unloading a train of small dump cars, 6 to 56 minutes. The large dump cars are operated by compressed air power furnished by the air pump of the locomotive, while the small dump cars are operated by hand.

The record day's work for one steam shovel was that of March 22, 1910, 4,823 cubic yards of rock (place measurement), or 8,395 tons. The highest daily record in the Central Division was on March 11, 1911, when 51 steam shovels and 2 cranes equipped with orange peel buckets excavated an aggregate of 79, 484 cubic yards, or 127,742 tons. During this day, 333 loaded trains and as many empty trains were run to and from the dumping grounds.

TOTAL AMOUNT OF EXCAVATION

The following was the estimated excavation required May 4, 1904, based on the plans for the lock canal:

	Cubic feet.
Atlantic Division	47,523,000
Central Division	106,417,000
Pacific Division	58,287,000
Total	212,227,000

Of this excavation, 180,423,874 cubic feet had been accomplished by Americans to October 1, 1912, leaving approximately 31,803,126 cubic feet remaining to be excavated.

The amount of material taken out by the Old and New Panama Canal Companies (French) was 78,146,960 cubic yards, of which it is estimated 29,908,000 cubic yards has been utilized in the adopted plan of Canal; making the total excavation for the Canal 242,135,000 cubic yards.

BREAKWATERS

Breakwaters are under construction at the Atlantic and Pacific entrances of the Canal. That in Limon Bay, or Colon harbor, extends into the bay from Toro Point, at an angle of 42 degrees and 53 minutes northward from a base line drawn from Toro Point to Colon light, and will be 10,500 feet in length, or 11,700 feet, including the shore connection, with a width at the top of fifteen feet and a height above mean sea level of ten feet. The width at the bottom will depend largely on the depth of water. It will contain approximately 2,840,000 cubic yards of rock, the core being formed of rock quarried on the mainland near Toro Point, armored with hard rock from Porto Bello. Work began on the breakwater in August, 1910, and on May 1, 1911, the fill had been extended 4,214 feet. The estimated cost is $5,500,000. A second breakwater has been proposed for Limon Bay, but this part of the project has not been formally acted upon. The purpose of the breakwaters is to convert Limon Bay into a safe anchorage, to protect shipping in the harbor of Colon, and vessels making the north entrance to the Canal, from the violent northers that are likely to prevail from October to January, and to reduce to a minimum the amount of silt that may be washed into the dredged channel.

The breakwater at the Pacific entrance will extend from Balboa to Naos Island, a distance of about 17,000 feet, or a little more than three miles. It will lie from 900 to 2,700 feet east of and for the greater part of the distance nearly parallel to the axis of the Canal prism; will vary from 20 to 40 feet in height above mean sea level, and will be from 50 to 3,000 feet wide at the top. It is estimated that it will contain about 18,000,000 cubic yards of earth and rock, all of which will be brought from Culebra Cut. It is constructed for a twofold purpose; first, to divert cross currents that would carry soft material from the shallow harbor of Panama into the Canal channel; second, to insure a more quiet harbor at Balboa. Work was begun on it in May, 1908. On May 1, 1911, it had been constructed for a distance of 13,000 feet.

CANAL FORCE, QUARTERS AND SUPPLIES

The Canal force is recruited and housed by the Quartermaster's Department, which has two general branches, labor and quarters, and material and supplies. Through the labor and quarters branch there have been brought to the Isthmus 43,432 laborers, of whom 11,797 came from Europe, 19,448 from Barbados, the balance from other islands in the West Indies and from Colombia. No recruiting is now required, the supply of labor on the Isthmus being ample.

In the month of August, 1912, there were approximately 45,000 employes on the Isthmus on the rolls of the Commission and of the Panama Railroad Company, about 5,000 of whom were Americans. There were actually at work on September 25, 1912, 35,861 men, 29,571 for the Commission, and 6,290 for the Panama Railroad Company. Of the 29,571 men working for the Commission, 4,166 were on the gold roll, which comprises those paid in United States currency, and 25,405 were on the silver roll, which comprises those paid on the basis of Panaman currency or its equivalent.

The gold force is made up of the officials, clerical force, construction men, and skilled artisans of the Isthmian Canal Commission and the Panama Railroad Company. Practically all of them are Americans. The silver force represents the unskilled laborers of the Commission and the Panama Railroad Company. Of these, about 4,500 are Europeans, mainly Spaniards, with a few Italians and other races. The remainder, about 25,000, are West Indians, about 3,700 of whom are employed as artisans, receiving 16, 20, and 25 cents, and a small number 32 and 44 cents, an hour. The standard rate of the West Indian laborer is 10 cents an hour, but a few of these doing work of an exceptional character are paid 16 and 20 cents. The larger part of the Spaniards are paid 20 cents an hour, and the rest 16 cents an hour.

The material and supply branch carries in eight general storehouses a stock of supplies for the Commission and Panama Railroad valued approximately at $4,500,000. About $12,000,000 worth of supplies are purchased annually, requiring the discharge of one steamer each day.

FOOD, CLOTHING AND OTHER NECESSARIES

The Canal and Panama Railroad forces are supplied with food, clothing and other necessaries through the Subsistence Department, which is divided into two branches—Commissary and Hotel. It does a business of about $7,500,000 per annum. The business done by the Commissary Department amounts to about $6,000,000 per annum, and that done by the hotel branch to about $1,500,000 per annum.

The Commissary system consists of 22 general stores in as many Canal Zone villages and camps along the relocated line of the Panama Railroad. It is estimated that with employes and their dependents, there are about 65,000 people supplied daily with food, clothing, and other necessaries. In addition to the retail stores, the following plants are operated at Cristobal: cold storage, ice making, bakery, coffee roasting, ice cream, laundry and packing department.

A supply train of 21 cars leaves Cristobal every morning at 4 a. m. It is composed of refrigerator cars containing ice, meats and other perishable articles, and ten containing other supplies. These are delivered at the stations along the line and distributed to the houses of employes by the Quartermaster's Department.

The hotel branch maintains the Hotel Tivoli at Ancon, and also 18 hotels along the line for white gold employes at which meals are served for thirty cents each. At these 18 hotels there are served monthly about 200,000 meals. There are sixteen messes for European laborers, who pay 40 cents per ration of three meals. There are served at these messes about 270,000 meals per month. There are also operated for the West Indian laborers fourteen kitchens, at which they are served a ration of three meals for 27 cents per ration. There are about 100,000 meals served monthly at these kitchens. The supplies for one month for the line hotels, messes and kitchens cost about $85,000; labor and other expenses about $17,500. The monthly receipts, exclusive of the revenue from the Hotel Tivoli, amount to about $105,000.

PANAMA CANAL STATISTICS

Length from deep water to deep water Bottom width of channel, maximum (feet)	1,000
Bottom width of channel, minimum, 9 miles, Culebra Cut (feet)	300
Locks, in pairs	12
Locks, usable length (feet)	1,000
Locks, usable width (feet)	110
Gatun Lake, area (square miles)	164
Gatun Lake, channel depth (feet)	85 to 45
Culebra Cut, channel depth (feet)	45
Excavation, estimated total (cubic yds.)	242,135,000
Excavation, amount accomplished by Americans September 30, 1912 (cubic yards)	180,423,874
Excavation by the French, useful to present Canal (cubic yards)	29,908,000
Total excavation by the French (cubic yards)	78,146,960
Excavation by the French, estimated value to Canal	$25,389,240
Value of all French property	$42,799,826
Concrete, total estimated for Canal (cubic yards)	5,000,000
Time of transit through completed Canal (hours)	10 to 12
Time of passage through locks (hours)	3
Relocated Panama Railroad, estimated cost	$ 9,000,000
Relocated Panama Railroad, length (miles)	47.1
Canal Zone, area (square miles)	448
Canal and Panama Railroad force actually at work May 1, 1912 (about)	35,000
Canal and Panama Railroad force, Americans (about)	5,000
Cost of Canal, estimated total	$375,000,000
Work begun by Americans	May 4, 1904
Date of completion, official	Jan. 1, 1915
Excavation remaining to be done Oct. 1, 1912, estimated (cubic yards)	31,803,126

VALUE OF THE $40,000,000 FRENCH PURCHASE

A careful official estimate has been made by the Canal Commission of the value to the Commission of the franchises, equipment, material, work done, and property of various kinds for which the United States paid the French Canal Company $40,000,000. It places the total value at $42,799,826, divided as follows:

Excavation, useful to the Canal, 29,708,000 cubic yards	$25,389,240.00
Panama Railroad Stock	9,644,320.00
Plant and material, used and sold for scrap	2,112,063.00
Buildings, used	2,054,203.00
Surveys, plans, maps and records	2,000,000.00
Land	1,000,000.00
Clearings, roads, etc	100,000.00
Ship channel in Panama Bay, four years' use	500,000.00
Total	$42,799,826.00

THE CANAL ZONE

The Canal Zone contains about 448 square miles. It begins at a point three marine miles from mean low water mark in each ocean, and extends for five miles on each side of the center line of the route of the Canal. It includes the group of islands in the Bay of Panama named Perico, Naos, Culebra, and Flamenco. The cities of Panama and Colon are excluded from the Zone, but the United States has the right to enforce sanitary ordinances in those cities, and to maintain public order in them in case the Republic of Panama should not be able, in the judgment of the United States, to do so.

Of the 448 square miles of Zone territory, the United States owns the larger portion, the exact amount of which is being determined by survey. Under the treaty with Panama, the United States has the right to acquire by purchase, or by the exercise of the right of eminent domain, any lands, buildings, water rights, or other properties necessary and convenient for the construction, maintenance, operation, sanitation, and protection of the Canal, and it can, therefore, at any time acquire the lands within the Zone boundaries which are owned by private persons.

Bird's-eye View of City of Panama.—New University of Panama on Left of Picture and Reservoir in Foreground.
—Photo by Underwood & Underwood, N. Y.

View Showing East Chamber, Lower and Middle Locks at Gatun, with Section of Rack Track on East Wall.
Photographed March 14, 1912.

Cylindrical Valve Machine, Motor, and Limit Switch. Electricity Is Used to Operate All Gates and Valves Along the Canal.

Miraflores Upper Locks, Center Wall Culvert, Showing Stoney Gate Castings in Place. Photographed June 23, 1912.

Rising Stem Gate Valve Machine. The Average Time of Filling and Emptying a Lock Will Be About Fifteen Minutes. The Valve System Will Furnish Perfect Control of the Water Flow.

Pedro Miguel Locks, Detail of Construction of Electric Towing Locomotive Rack Track. All Vessels Will Be Towed by Electricity Through the Canal.

Front Tower, Range 5-6, Atlantic Division. These Range Lights Form an Important Feature of the Navigation Facilities of the Canal System.

Front Tower, Range 9-11, Pacific Entrance Looking Southeast. Photographed November 7, 1911, and Showing the Reinforced Concrete Construction of the Lighthouses.

Rear Tower, Range 9-11, Pacific Entrance, Looking Northwe
Will Make the Navigation of the Canal

Gatun Spillway Looking Southwest, Showing Downstream Face of Ogee Dam, as it Appeared June 6, 1912.

South End of Naos Island Dump, 4,000 Feet from Island. Center at "A" is 75 Feet from Track and 25 Feet Above the Original Bottom. Elevation of Trestle, +14. Photographed December, 1911.

Stripping Cocoli Hill Adjacent to Canal Prism. Photographed March 21, 1912, and Showing the Method of Hydraulic Excavation.

Miraflores Lower Locks, Looking East. Electrical Conduit and Floor Culvert at Upper End of East Lock, as They Appeared January 19, 1912.

Gatun Upper Locks, West Chamber, Looking North, Showing Upper Guard Gates, Operating Gates, Intermediate Gates, and Safety Gates in Process of Construction, June 7, 1912.

Gatun Upper Locks, East Chamber, Looking North from Forebay, Showing Upper Guard Gates and Emergency Dam Sill, July 7, 1912. The Lock Gates Are Steel Structures 7 Feet Thick and Weigh from 300 to 600 Tons Each.

Gatun Upper Locks, Looking North from Lighthouse, as They Appeared July 2, 1912. The Three Pairs of Locks at Gatun Have a Combined Lift of 85 feet.

Pedro Miguel Locks, South End of East Chamber, Showing Construction of Safety and Lower Gates as They Appeared June 3, 1912. See Model of These Locks on Page 17.

Gatun Lower Locks Looking South fom Cofferdam, Showing West Chambers of Upper and Middle Locks, as They Appeared on November 9, 1911.

Gatun Upper Locks, Miter Gate Moving Machine, Structural Steel Girders for Towing Locomotive Track Supports in Foreground. This Photograph Was Taken in June, 1912.

Gatun Locks Forebay, East Side Looking North, Showing Flaring Approach Wall. Photographed June 7, 1912.

Gatun Dam, West Section of Dam Looking West, Showing Progress of Hydraulic Fill, June 12, 1912. This Great Dam Is Nearly 1½ Miles Long and About ½ Mile Wide at Its Base.

Culebra Cut, South End, Looking South from Bridge 57½ and Showing the Partly Completed Anchorage Basin North of Pedro Miguel Lock. Train Is on Completed Bottom of Canal, Elevation +40. Photographed June, 1912.

Gatun Spillway Looking East Toward Locks, Showing Up and Down Stream Faces of Ogee Dam, June 6, 1912. The Spillway Will Be Used to Regulate the Water Level of Gatun Lake.

Culebra Cut Looking North from Las Cascadas. All Trains Are Standing on the Bottom of the Cut, Elevation +40.
Photographed May, 1912.

Culebra Cut Looking North from Bridge 57½, Near Paraiso. The Train on the Left, Just Beyond the Trestle Bridge, Is on the Completed Bottom of the Canal, Elevation +40. Photographed June, 1912.

Culebra Cut, Looking South from Empire Suspension Bridge. The Group of Well Drills in the Middle of the Canal Is About 27 Feet Above the Bottom, or at Elevation +67, Photographed May, 1912.

Pedro Miguel Locks. Bird's-eye View from Hill on East Bank. The Photograph Was Taken July 28, 1912.

Culebra Cut, Culebra. Break in East Bank of Canal. Amount of Material Involved, 320,000 Cubic Yards. The Train Shown in Foreground Is About 35 Feet Above the Bottom, or at Elevation +75. Photographed February 11, 1912.

Pedro Miguel Locks. Bird's-eye View of North Approach Wall from Hill at East End, as It Appeared July 28, 1912.

Miraflores Locks Looking North, as They Appeared June 21, 1912. See Description of Locks on pages 14 to 18.

Miraflores Locks, West Chamber, Looking South. Photographed June 23, 1912. Each Lock Has a Usable Length of 1,000 Feet and Width of 110 Feet.

Pedro Miguel Locks, Looking South. West Forebay, with Emergency Dam Sill. Photographed June 5, 1912. See Model of Locks on Page 17.

Slide of Stratified Rock, West Bank of Canal, Culebra-on-the-Dump, Looking Toward Culebra. Slide Involves About 1,000,000 Cubic Yards and Moved About 3 Feet Per Day on a Slope of 1 Vertical to 7 Horizontal. The Train Is Standing at Elevation +95. Photographed February, 1912.

Miraflores Upper Locks. General View Looking North from Lower West Bank, Showing Cylindrical Valves
Photographed July 25, 1911.

Pedro Miguel Locks, North End of West Chamber Showing Construction of Upper Guard Gates and Upper Gates, as They Appeared June 5, 1912.

Balboa—Lumber Dock of Reinforced Concrete, Looking Northeast. June, 1912. This Pacific Port at the Southwestern End of the Canal Will Benefit Largely from Its Construction.

Slide in East Bank of Canal Near Cucaracha, June, 1912. This Illustrates One of the Difficulties with Which the Engineers and Construction Department Have Had to Contend.

Culebra Cut Looking South from Bend in East Bank Near Gamboa. The Train and Shovel Are Standing on the Bottom of the Cut. The Water in the Drainage Channel Is About 10 Feet Below the Bottom of the Canal, or at Elevation +30. Photographed June, 1912.

Empire-Chorrera 16-Foot Macadam Road Under Construction with Zone Prison Labor, as It Appeared August 29, 1912

Pedro Miguel Locks Looking North, Showing Upper Guard Gates, East Chamber Forebay, and Construction of Approach Wall. The Scene Was Photographed March 28, 1912.

Steam Shovel 218 Buried Under Fall of Rock, West Side of Canal, Near Las Cascadas. This Shovel Was Working on the Bottom of the Canal When Destroyed, May 31, 1912.

Miraflores Upper Locks, Showing Stoney Gate Valve Frames in Position in South End of West Wall.
Photographed November 8, 1911.

West Breakwater, Looking Seaward from Toro Point, Showing Dredge at Work Placing Rock on Face of Breakwater. Photographed June, 1912.

Gatun Lower Lock, East Chamber, Looking North, Showing Temporary Cofferdam at Extreme End of Lock Chamber. The View Was Taken June 12, 1912.

Gatun Locks. General View Looking Southwest, Showing North End of the Locks, with Temporary Cofferdam in Place. Photographed July 2, 1912.

Heated Material on the West Side of the Canal, 350 Yards North of Culebra y. Extent of Heated Material 500' x 25' x 20'. Photographed February 16, 1912.

Culebra Cut, Looking North from Empire Suspension Bridge. The Nearest Shovel Shown, in the Lowest Cut, Is Working About 12 Feet Above the Bottom, or at Elevation +52. The Photograph Was Taken in May, 1912.

Culebra Cut Looking North from Cunette. The Two Shovels Shown in the Foreground Are Working on the Bottom, Elevation +40. The Water Standing in the Center Drainage Channel Is About 6 Feet Below the Bottom, Elevation +34. Photographed in June, 1912.

South End of Naos Island Dump, 4,000 Feet from Island. Center at "A" Is 80 Feet from Track and 25 Feet Above the Original Bottom. Elevation of Trestle, +14. Photographed in December, 1911.

Miraflores Lower Locks. Slide Back of West Wall, Looking South, as It Appeared March 21, 1912.

Miraflores Locks. Sinking Caissons for Foundation of North Approach Wall, Looking North, June 14, 1912.

Columnar Structure in Hardened Flows of Mud Lava. This Jointing Afforded Passages for Seepage Water Which Tended Largely to Promote a Large Slide Just North of La Pita.

"A" Fault Plane. "B" Crushed and Sheared Zone of Rock. "C" Stronger Rocks, Beds of Limy Sandstone. This Fault Was the Chief Cause of the Big Slide on the West (Opposite) Side of the Canal Near Lirio.

Front Lighthouse Tower, Range 1-2 Gatun Lake Section, on South Middle Approach Wall of Gatun Locks.

Rear Tower, Lighthouse, Range 3-4. (9-11) Pacific Entrance of the Canal at Low Tide. Front Tower.

Gatun Locks and Dam, Looking West from Water Tower, Showing South Center Approach Wall and Forebay of Gatun Locks, with Dam and Spillway in the Distance. Photographed June, 1912.

Gatun Spillway Looking Southwest, Showing Downstream Face of Ogee Dam, as it Appeared June 6, 1912.

Culebra Cut Looking South from Cunette. The Two Shovels Shown in the Foreground Are Working on the Bottom of the Canal, Elevation +40. Photographed in May, 1912.

ulebra Cut, Looking North from a Point South of Contractor's Hill, Showing Quiescent State of the Great Cucaracha Slide on Right Bank. Photographed May, 1912.

Channel Excavated at San Pablo During Dry Season, 1912. This Channel Is Completed and Is 800 Feet Wide, With Bottom Elevation at +40.

Embankment of Old Panama Railroad Excavated Down to +35 in April and May, 1912.

General View of Balboa Terminal Site, Looking North, June, 1912.

Dredge "Corozal" in Channel Near Station 2210 of the Canal Operations. Photographed in June, 1912.

Mandingo Stockade for Zone Convicts Engaged in Road Building. The Photograph Was Made in August, 1912.